Fun with Shapes

FUN WITH SQUARES

Bert Wilberforce

PowerKiDS press
New York

Let's look for squares!

The grass looks like a square.

The letter looks like a square.

The watch looks like a square.

The soap looks like a square.

The bag looks like a square.

The stamp looks like a square.

The brownie looks like a square.

The cracker looks like a square.

The gift looks like a square.

Do you see squares?

Published in 2023 by The Rosen Publishing Group, Inc.
29 East 21st Street, New York, NY 10010

Copyright © 2023 by The Rosen Publishing Group, Inc.

All rights reserved. No part of this book may be reproduced in any form without permission in writing from the publisher, except by a reviewer.

First Edition

Editor: Therese Shea
Book Design: Rachel Rising

Photo Credits: Cover, p. 1 Nickolastock/Shutterstock.com; p. 3 Mega Pixel/Shutterstock.com; p. 5 Sarunyu_foto/Shutterstock.com; p. 7 slava17/Shutterstock.com; p. 9 Smirnof/Shutterstock.com; p. 11 Sisacorn/Shutterstock.com; p. 13 suksawad/Shutterstock.com; p. 15 Tony Baggett/Shutterstock.com; p. 17 MaraZe/Shutterstock.com; p. 19 mayakova/Shutterstock.com; p. 21 Sarah2/Shutterstock.com; p. 23 Creatus/Shutterstock.com.

Library of Congress Cataloging-in-Publication Data

Names: Wilberforce, Bert, author.
Title: Fun with squares / Bert Wilberforce.
Description: New York : PowerKids Press, [2023] | Series: Fun with shapes
Identifiers: LCCN 2021046336 (print) | LCCN 2021046337 (ebook) | ISBN
  9781538385555 (library binding) | ISBN 9781538385531 (paperback) | ISBN
  9781538385548 (set) | ISBN 9781538385562 (ebook)
Subjects: LCSH: Square--Juvenile literature.
Classification: LCC QA482 .W585 2023 (print) | LCC QA482 (ebook) | DDC
  516/.154--dc23/eng/20211117
LC record available at https://lccn.loc.gov/2021046336
LC ebook record available at https://lccn.loc.gov/2021046337

Manufactured in the United States of America

Some of the images in this book illustrate individuals who are models. The depictions do not imply actual situations or events.

CPSIA Compliance Information: Batch #CSPK23. For further information contact Rosen Publishing, New York, New York at 1-800-237-9932.